GOODS OF THE MIND, LLC

Competitive Mathematics Series

for

Gifted Students in Grades 1 and 2

PRACTICE OPERATIONS

Cleo Borac, M. Sc.
Silviu Borac, Ph. D.

This edition published in 2013 in the United States of America.

Editing and proofreading: David Borac, B.Mus.
Technical support: Andrei T. Borac, B.A., PBK

Send all inquiries to:

Goods of the Mind, LLC
1138 Grand Teton Dr.
Pacifica
CA, 94044

Competitive Mathematics Series for Gifted Students
Level I (Grades 1 and 2)
Practice Operations

Contents

FOREWORD

The goal of these booklets is to provide a problem solving training ground starting from the earliest years of a student's mathematical development.

In our experience, we have found that teaching how to solve problems should focus not only on finding correct answers but also on finding better solution strategies. While the correct answer to a problem can typically be obtained in several different ways, not all these ways are equally useful for learning how to solve problems.

The most basic strategy is *brute force*. For example, if a problem asks for the number of ways Lila and Dina can sit on a bench, it is easy to write down all the possibilities: Dina, Lila and Lila, Dina. We arrive at this solution by performing all the possible actions allowed by the problem, leaving nothing to the imagination. For this last reason, this approach is called brute force.

Obviously, if we had to figure out the number of ways 30 people could stand in a line, then brute force would not be as practical, as it would take a prohibitively long time to apply.

Using brute force to obtain the correct answer for a simpler problem is not necessarily a useful learning experience for solving a similar problem that is more complex. Moreover, solving problems in a quantitative manner, assuming that the student can transfer simple strategies to similar but more complex problems, is not an efficient way of learning problem solving.

From this simple example, we see that the goal of *practicing* problem solving is different from the goal of problem solving. While the goal of problem solving is to obtain a correct answer, the goal of practicing problem solving is to acquire the ability to develop strategies, generate ideas, and combine approaches that are powerful enough to solve the problem at hand as well as future similar problems.

While brute force is not a useless strategy, it is not a key that opens every door. Nevertheless, there are problems where brute force can be a useful tool. For instance, brute force can be used as a first step in solving a complex problem: a smaller scale example can be approached using brute force to help the problem solver understand the mechanics of the problem and generate ideas for solving the larger case.

All too often, we encounter students who can quickly solve simple problems by applying brute force and who become frustrated when the solving methods they have been employing successfully for years become inefficient once problems increase in complexity. Often, neither the student nor the parent has a clear understanding of why the student has stagnated at a certain level. When the only arrows in the quiver are guess-and-check and brute force, the ability to take down larger game is limited.

Our series of books aims to address this tendency to continue on the beaten path - which usually generates so much praise for the gifted student in the early years of schooling - by offering a challenging set of questions meant to build up an understanding of the problem solving process. Solving problems should never be easy! To be useful, to represent actual training, problem solving should be challenging. There should always be a sense of difficulty, otherwise there is no elation upon finding the solution.

Indeed, practicing problem solving is important and useful only as a means of learning how to develop better strategies. We must constantly learn and invent new strategies while questioning the limitations of the strategies we are using. Obtaining the correct answer is only the natural outcome of having applied a strategy that worked for a particular problem in the time available to solve it. Obtaining the wrong answer is not necessarily a bad outcome; it provides insight into the fallacies of the method used or into the errors of execution that may have occured. As long as students manifest an interest in figuring out strategies, the process of problem solving should be rewarding in itself.

Sitting and thinking in a focused manner is difficult to train, particularly since the modern lifestyle is not conducive to adopting open-ended activities. This is why we would like to encourage parents to pull back from a quantitative approach to mathematical education based on repetition, number of completed pages, and the number of correct answers. Instead, open up the time boundaries that are dedicated to math, adopt math as a game played in the family, initiate a math dialogue, and let the student take his or her time to think up clever solutions.

Figuring out strategies is much more of a game than the mechanical repetition of stepwise problem solving recipes that textbooks so profusely provide, in order to "make math easy." Mathematics is not meant to be easy; it is meant to be interesting.

Solving a problem in different ways is a good way of comparing the merits of each method - another reason for not making the correct answer the primary goal of the activity. Which method is more labor intensive, takes more time or is more prone to execution errors? These are questions that must be part of the problem solving process.

In the end, it is not the quantity of problems solved, the level of theory absorbed, or the number of solutions offered in ready-made form by so many courses and camps, but the willingness to ask questions, understand and explore limitations, and derive new information from scratch, that are the cornerstones of a sound training for problem solvers.

These booklets are not a complete guide to the problem solving universe, but they are meant to help parents and educators work in the direction that, aside from being the most efficient, is the more interesting and rewarding one.

The series is designed for mathematically gifted students. Each book addresses an age range as some students will be ready for this content earlier, others later. If a topic seems too difficult, simply try it again in a couple of months.

OPERATIONS WITH INTEGERS

Notes to parents: The student has to understand the following aspects of operations:

1. The operation has a *syntax*, a way of writing it down so people and machines can understand what it means. This syntax can be changed. Civilizations have specified operations with numbers in different ways throughout history.

2. The *order of operations* is a construction that allows us to tell people and machines which operation should be the next in line to be performed. It is generally part of the *syntax* and it is just a convention. Students should observe several such conventions in order to get a feeling that syntax, as learned in school, is not cast in stone. We include a number of examples that use ad-hoc syntax which the student has to figure out.

3. While numbers are *abstract*, their *writing* is concrete. The symbols used to represent numbers, however, have changed through time. For example, the Arabic numeral 4 is different from the Roman numeral IV but both represent the same abstract notion. There is only a single number 4.

This book has several different goals:

- To help the student think flexibly, not mechanically, about operations.

- To help the student manipulate operations with numbers by using a variety of techniques in addition to the algorithms typically found in school textbooks.

- To develop a solid number sense by indicating what is changeable and what is not changeable in our arithmetic.

This section uses operations for both the first and the second grade level. Some students will be more ready, others less ready, for the entire content. If necessary, students can skip the more difficult problems at the first reading and come back to them a few months later.

PRACTICE ONE

Do not use a calculator for any of the problems!

Exercise 1

Compute the sum:

$$16 + 0 =$$

Exercise 2

True or false?

$$102 - 0 = 102 + 50 - 50$$

Exercise 3

True or false?

$$102 + 0 = 102 - 50 + 50$$

Exercise 4

True or false?

$$5 + 2 - 2 = 2 - 2 + 5$$

Exercise 5

Dina computed the following:

$$100 + 11 - 11 + 12 - 12 + 13 - 13 =$$

$$100 - 11 + 11 - 12 + 12 - 13 + 13 =$$

She noticed something and found an explanation! She told Lila about it. What do you think Dina's explanation was?

Exercise 6

Lila computed the following very quickly:

$$100 + 99 + 98 + 97 + 96 + 95 + 94 - 95 - 96 - 97 - 98 - 99 - 100 =$$

How do you think she was able to do it so quickly?

Exercise 7

What is the value of the number covered by the clover?

$$77 - \clubsuit = 34$$

Exercise 8

What is the value of the number covered by the diamond?

$$108 - \diamondsuit = 99$$

Exercise 9

Dina performed these additions and noticed something interesting. What did Dina notice?

$$123 + 321 \ =$$
$$342 + 243 \ =$$
$$241 + 142 \ =$$
$$721 + 127 \ =$$

Exercise 10

Place the numbers 15, 12, and 11 in the squares to make the following equality true:

$$\square \quad + \quad \square \quad - \quad \square \quad = \quad \textbf{16}$$

Exercise 11

Dina's calculator has a different way of understanding computations. To add two numbers, Dina has to enter + and then the two numbers she wants to add, like this:

$+\ \ 4\ \ 5$

When she presses **Enter**, the calculator displays the result: 9.
If she wants to subtract 4 from 5, Dina has to enter:

$-\ \ 5\ \ 4$

The calculator scans the operations from left to right. It performs the operations that are directly followed by the numbers they apply to and replaces the whole group with the result. It then repeats this until only one result remains. For example:

If Dina wants to add 5 and 4 and then subtract 2 from the sum, which of the following is she going to have to enter?

(A) + 4 5 − 2

(B) + − 5 4 2

(C) − + 4 5 2

Exercise 12

More practice with Dina's calculator:

(a) + 7 8

(b) + + 7 8 2

(c) − + 7 8 2

(d) + 5 3

(e) − 5 3

(f) × 5 2 2

(g) ÷ × 5 2 2

(h) ÷ + 5 5 2

(i) × − 5 5 4

(j) + + + 3 3 3 3

(k) − − − 4 1 1 1

(l) − + 4 3 2

(m) − + 1 9 1

Note: This syntax is called the *Polish notation* and was invented by Jan Lukasiewicz in 1924.

Exercise 13

Fill in the missing values.

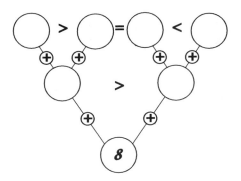

Exercise 14

By using parentheses, one can tell a calculator to perform the operation in the parentheses before the other operations. For example,

$$2 \times (5 - 2)$$

will instruct the calculator to do the subtraction first, then multiply the result by 2.

After performing the subtraction, the parentheses are no longer needed, and $(5 - 2)$ is replaced by 3, like this:

$$2 \times 3$$

Dina and Lila have entered the following operations in their calculators. Each calculator performs one operation at a time. Fill in the blanks with the numbers each calculator used for each step.

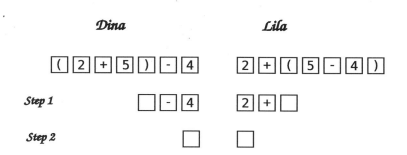

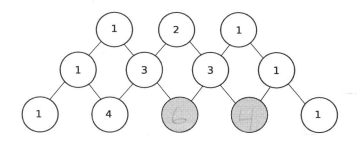

Exercise 15

Find a pattern and fill in the missing values:

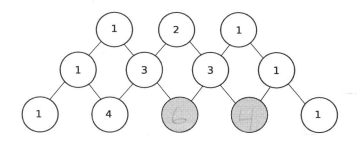

Exercise 16

Find the number in the grey circle:

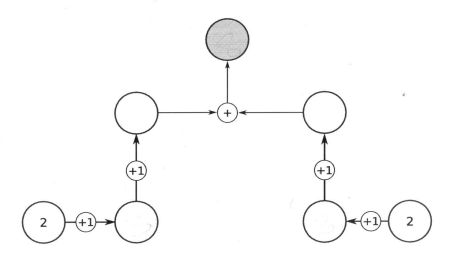

Exercise 17

Place the results of the operations in the boxes such that identical results go in the same box. How many boxes have remained empty?

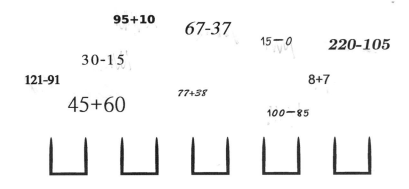

USING NEGATIVE INTEGERS AND PARENTHESES

A succession of additions and subtractions can be thought of as an **algebraic sum**: a sum of positive and negative numbers. By doing this, we can get rid of subtractions and one of the unpleasant properties associated with them: lack of commutativity.

Subtraction is not commutative:

$$7 - 4 \neq 4 - 7$$

However, addition is:

$$7 - 4 = -4 + 7$$

The advantage of introducing negative numbers is that they allow us to think of subtractions as additions. Now, the terms may be moved *provided we move them together with their sign:*

$$4 - 3 + 5 - 4 + 6 - 5 = 4 + 5 + 6 - 3 - 4 - 5 = 4 - 4 + 5 - 5 + 6 - 3$$

This distinction is often a source of confusion for students. They ask "How can you swap the terms? Didn't you say subtraction was not commutative?" Notice that we haven't swapped the terms around the minus sign! We have moved the term together with its sign:

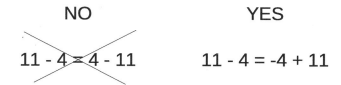

NO	YES
11 - 4 = 4 - 11	11 - 4 = -4 + 11

Operations enclosed in parentheses have to be performed first. The parentheses are then replaced by the result (but not the operator in front of the leftmost parenthesis!)

If you are a machine:

1. Scan the expression *from left to right.*

2. Perform the operations in the *innermost* parentheses.

3. Replace the parentheses with the result.

4. Go back to step i. Stop when there is a single number left.

If you are *not* a machine, look at the expression carefully. Who knows, maybe there are some details that will help you find the result without doing so many computations!

Look at these two ways of processing operations:

3+4+5+6-2-3-4-5   3+4+5+6-2-3-4-5

3+4+5+6-(2+3+4+5)   6-5+5-4+4-3+3-2

18-(14)   1+1+1+1

4   4

Both of the procedures are better than performing each operation in turn from left to right. The procedure on the right hand side, however, is more efficient than the procedure on the left.

The procedure on the left illustrates that instead of subtracting a set of numbers, we can *subtract their sum*. Instead of working to remove parentheses, *we created them* because they helped us plan the order of our operations.

PRACTICE TWO

Do not use a calculator for any of the problems!

Exercise 1

True or false?

$$1 - 2 + 1 = 1 + 1 - 2$$

Exercise 2

True or false?

$$3 - 4 = 4 - 3$$

Exercise 3

True or false?

$$3 - 4 = -4 + 3$$

Exercise 4

Dina and Lila play a game on the number line. If Dina says a positive number, Lila will move that number of steps to the right. If Dina says a negative number, Lila will move that number of steps to the left. When Dina does not say anything, Lila stays put. Dina and Lila are facing each other. Dina says:

$$-2, 3, 1, -4, 2, -1, 3, -2, 1, 1, -4$$

Is Lila now to Dina's left or to Dina's right? How many steps away from Dina is Lila now?

Exercise 5

Lila must place the operators $+$ and $-$ in the empty squares so that the following equality is true. Can you help her?

Exercise 6

Compute efficiently:

(a) $22 - 5 + 1 - 8 =$

(b) $101 + 59 - 100 =$

(c) $119 + 1 - 49 - 1 =$

(d) $77 - 2 - 3 - 4 - 5 - 6 =$

(e) $299 + (1 - 299) =$

Exercise 7

True or false?

(a) $11 - 4 - 5 - 6 - 7 + 99 = 110 - (4 + 5 + 6 + 7)$

(b) $20 - 30 + 10 = 20 + 10 - 30$

(c) $2 - 3 + 4 - 5 + 6 = 6 - 5 + 4 - 3 + 2$

(d) $(13 + 12) - (11 + 14) = 12 - 11 + 13 - 14$

Exercise 8

True or false?

(a) $23 + 25 + 27 = 28 - 1 + 26 - 1 + 24 - 1$

(b) $16 + (1 - 16) = (16 - 16) + 1$

(c) $11 - (5 + 6) = 11 - 5 + 6$

(d) $5 + 15 + 25 - 20 - 10 = 5 + (15 - 10) + (25 - 20)$

Exercise 9

Lila has to place parentheses to obtain the smallest possible result:

$$43 \quad - \quad 3 \quad + \quad 15 \quad - \quad 7$$

Exercise 10

Dina has to place parentheses to obtain the result shown:

$$49 \quad - \quad 8 \quad - \quad 41 \quad + \quad 37 \quad = \quad 45$$

Exercise 11

Compute the result:

$$2 + 4 + 6 + 8 + 10 - 1 - 3 - 5 - 7 - 9 =$$

Exercise 12

Dina has to place parentheses in the following expression so that the computation is correct:

$$40 \quad + \quad 8 \quad \div \quad 8 \quad = \quad 6$$

Exercise 13

Lila has to place parentheses in the following expression so that the computation is correct:

$$9 \quad + \quad 8 \quad \times \quad 3 \quad = \quad 51$$

Exercise 14

Place parentheses in the following expression so that the computation is correct:

$$8 \; + \; 10 \; \times \; 2 \; = \; 36$$

Exercise 15

Place parentheses in the following expression so that the computation is correct:

$$4 \; + \; 5 \; \times \; 9 \; - \; 3 \; = \; 54$$

Exercise 16

Without calculating, can you tell whether each line is true or false?

$$10 - 1 + 11 - 1 \; = \; 10 + 11 - 2$$
$$8 - 3 \; = \; -3 + 8$$
$$8 - 3 \; = \; 3 - 8$$
$$5 + 3 - 3 \; = \; 8 + 5 - 8$$
$$11 - 12 + 12 - 1 \; = \; 11 - 1$$

Exercise 17

True or false?

$$2 - 2 + 3 - 3 + 4 - 4 + 5 - 5 = -2 + 2 - 3 + 3 - 4 + 4 - 4 + 5$$

Exercise 18

True or false?

$$5+6+7+8+9-5-6-7-8-9 = 5+6+7+8+9-(5+6+7+8+9)$$

LARGE NUMBERS OF OPERANDS

At this age level, students will be able to count objects they see. For example, they will be able to count 6 numbers in this list: $\{1, 2, 3, 4, 5, 6\}$.

Let us imagine that the list is much longer, comprising the numbers from 1 to 40. We use the following notation to write the list:

$$1, \ 2, \ 3, \ 4, \ \ldots, \ 40$$

When asked how many numbers there are in the list, some students will answer 5 (which is the number of numbers they see) and others will answer 40 (which is the correct answer).

Students begin to understand that "there are more numbers behind the dots" at various ages, ranging from the elementary to the upper middle school grades. Precocious kids are able to work with this abstraction as early as first or second grade.

In this book, we try to provide exercises that familiarize the student with operating on large sets of numbers. Educators should take the student's current abilities into account when using this material and come back to it later if the student finds it too difficult.

Experiment

In the following experiment, we do not move or remove any of the cubes. We only hide some of them.

Look at this picture. How many cubes are there?

Look at this picture. How many cubes are there behind the transparent screen? How many cubes are there in total?

Look at this picture. How many cubes are there behind the opaque screen? How many cubes are there in total?

Look at this picture. How many cubes have been replaced by dots? How many cubes are there in total?

PRACTICE THREE

Exercise 1

In the following sequence, dots have been used to replace some numbers that we do not want to write. Since the numbers follow a pattern, we can imagine the missing numbers without having to write them. Make a list of the missing numbers:

$$\{2, 4, 6, 8, 10, \ldots, 20\}$$

(a) How many numbers are written out?

(b) How many numbers have been replaced by dots?

(c) How many numbers are there in total (written and unwritten)?

Exercise 2

Make a list of the missing numbers:

$$\{2, 4, 6, 8, 10, \ldots, 40\}$$

(a) How many numbers are written out?

(b) How many numbers have been replaced by dots?

(c) How many numbers are there in total (written and unwritten)?

Exercise 3

Do not make a list of the missing numbers! There are too many of them. Based on your solutions to the previous two exercises, can you find out how many numbers have not been written out?

$$\{2, 4, 6, 8, 10, \ldots, 400\}$$

How many numbers are there in total in this sequence (written and unwritten)?

Exercise 4

Dina has to answer some questions about the following expression:

$$1 + 2 + 3 + \cdots + 40$$

1. How many numbers are added together?
2. How many of these numbers are odd?
3. How many $+$ operators are there in total?
4. Will the result be even or odd? (Answer without calculating the result.)

Exercise 5

How many squares make a triangle?

$$\square + \square + \square + \square = \triangle + \square$$

Exercise 6

How many squares make a star?

$$\underbrace{\square + \square + \cdots + \square}_{101 \text{ squares}} = \text{☆} + \square$$

Exercise 7

How many squares make a circle?

$$\square + \square + \cdots + \square = \bigcirc + \bigcirc + \bigcirc$$

$$\longleftarrow \text{66 squares} \longrightarrow$$

Exercise 8

Dina has to count the number of operations in the following sum:

$$1 + 2 + 3 + 4 + 5 + 6 + 7 + 8 + 9 + 10$$

What has Dina found?

Exercise 9

Lila has to count the number of operations in the following sum:

$$1 + 2 + 3 + 4 + 5 + \cdots + 100$$

If Lila knows what Dina found in the previous problem, can she find an answer without writing out all the numbers?

Exercise 10

Compute efficiently:

$$2 + 4 + 6 + 8 + 10 + \cdots + 50 - 1 - 3 - 5 - 7 - 9 - \cdots - 49 =$$

Exercise 11

How many numbers are there in the list?

$$\{11, \ 22, \cdots, 99, \ 111, \ 222, \cdots, 999\}$$

Exercise 12

How many numbers are there in the list?

$$\{1, \ 11, \ 111, \ 1111, \cdots, 111111111111111\}$$

Exercise 13

How many numbers in the list are even?

$$\{0, \ 1, \ 2, \ 3, \cdots, 100\}$$

Exercise 14

How many towers are there?

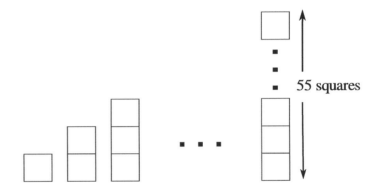

Exercise 15

How many circles are shaded?

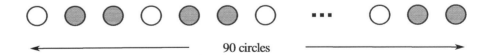

Exercise 16

How many digits have been used to write the following list:

$$\{1, \ 22, \ 333, \cdots, 7777777, \cdots, 333, \ 22, \ 1\}$$

Exercise 17

Lila and Dina are playing a game of dots. They use a ruled sheet of paper. Dina starts by drawing a dot on the first line. Lila draws two dots on the second line. Dina draws three dots on the third line. They continue on until one of them draws 47 dots on a line. Which line is it? Who draws the 47 dots, Lila or Dina?

UNDERSTANDING EQUATIONS

It is not too early to familiarize the student with equations.
First of all, start with the obvious observation:

$$4 = 4$$

Now let us say the square represents some number. Also, the square *always* represents the same number. The fact that the square never changes its value, enables us to say that:

$$\square \;=\; \square$$

The equality is an *identity* - it states that the square on the left is identical to the square on the right. This will be true if we choose the square to be the number 5 and it will also be true if we choose the square to be the letter Q.

The following, however, is not an identity anymore:

$$\square \;+\; \bigcirc \;=\; \square \;+\; \mathbf{3}$$

The square can still be anything but the circle must be the number 3.

This last situation is called an *equation*. It is an equality that is true *only for certain values* of the symbols used.

PRACTICE FOUR

Do not use a calculator for any of the problems!

Exercise 1

Which number is hiding behind the square?

$$\square \;+\; \square \;=\; \boxed{40}$$

Exercise 2

Which number is hiding behind the circle?

$$\bigcirc \;+\; \bigcirc \;=\; 22$$

Exercise 3

Fill in the circles with appropriate integer numbers:

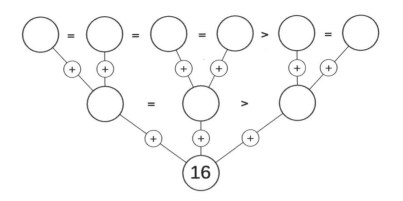

Exercise 4

Find K:

$$61 + 59 = K + K + K$$

Exercise 5

Place the same number in each empty circle:

Exercise 6

Place in the circles numbers formed using the same digit:

Exercise 7

Place operators $(+, -, \times, \div)$ within the boxes. No parentheses are needed.

Exercise 8

In each of the following, which number does the letter represent?

$$A + A + A = 15$$

$$B + B + B + B - B = 15$$

$$C + C + C + C + C = 15$$

$$C + C + C + C + C + C - C = 15$$

$$D + D - D + D - D + D - D + D - D + D - D = 15$$

Exercise 9

The triangle, the circle, and the square have different weights. Which of the following would balance the scale on the right?

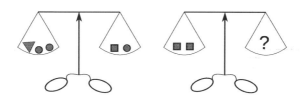

(A) one triangle and one circle

(B) one triangle and three circles

(C) two triangles and one circle

(D) two triangles and two circles

Exercise 10

Amira is out shopping for Hallowe'en treats. She wants to buy some Real Fruit Eyeballs and some Fruit Jelly Smellies. Three packs of Eyeballs and two packs of Smellies cost 9 dollars. Two packs of Eyeballs and three packs of Smellies cost 11 dollars. Amira has 12 dollars. Does she have enough money to buy three packs of Eyeballs and three packs of Smellies?

MISCELLANEOUS PRACTICE

Do not use a calculator for any of the problems!

Exercise 1

Which number does the circle represent?

$$\bigcirc + \bigcirc = \stackrel{\displaystyle \star}{}$$

$$\stackrel{\displaystyle \star}{} + \stackrel{\displaystyle \star}{} = 136$$

Exercise 2

Replace the question mark with a number so that the operations are correct:

$$\bigcirc + \stackrel{\displaystyle \star}{} = 16$$

$$\bigcirc - \stackrel{\displaystyle \star}{} = 4$$

$$\bigcirc + \bigcirc = \;?$$

Exercise 3

If two positive integers have an odd difference, is their sum:

(**A**) always even?

(**B**) always odd?

(**C**) sometimes even and sometimes odd?

Exercise 4

There are 100 plus signs in the following operation. What is the result of the additions?

$$1+1+1+\cdots+1 = ?$$

Exercise 5

Roman numerals time! Perform the following operations and write the answer in both Roman and Arabic numerals:

(a) $I + I =$

(b) $II + II =$

(c) $III + III =$

(d) $IV + IV =$

(e) $V + V =$

(f) $XX + XX =$

(g) $XXX + XXX =$

(h) $LX + LX =$

(i) $L + L =$

(j) $C + C =$

(k) $CC + CC =$

(l) $CCC + CCC =$

(m) $CD + CD =$

(n) $CD + D =$

Exercise 6

Each group of four numbers in the sequence follow the same pattern. Which numbers belong in the empty circles?

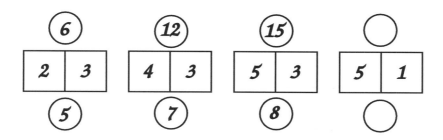

Exercise 7

In the figure, shaded circles of the same color hide the same number. Which number does the white circle hide?

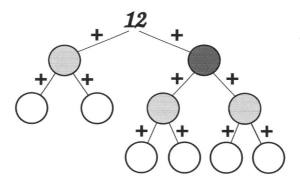

Exercise 8

Compute:

$$4 + 3 - 4 + 5 - 6 + 6 - 7 + 8 - 9 =$$

Exercise 9

Dina had 30 party balloons. During the party, 11 balloons popped. Afterwards, Lila gave Dina as many balloons as had popped. How many party balloons did Dina have then?

Exercise 10

The table in the figure has 10 rows and 11 columns. How many grey cells are there in the table? How many grey cells should we color white in order to have an equal number of white and grey cells?

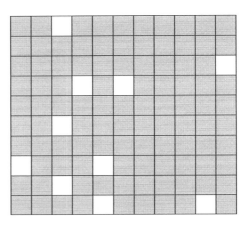

Exercise 11

In the lists below, some digits, not necessarily identical, have been hidden by a ♠. In each list, there are two different 3-digit numbers. Find the lists in which it is possible to tell which number is smaller.

(A) 9♠7, ♠87

(B) 33♠, 31♠

(C) 4♠3, 5♠♠

(D) 9♠♠, 9♠0

Exercise 12

Dina has 20 magnets and Lila has 11 magnets. How many magnets could Amira have if she has fewer than Dina and more than Lila? Check all that apply.

(A) 9

(B) 10

(C) 13

(D) 19

Exercise 13

How many 3-digit numbers have a digit sum of 2?

Exercise 14

A 100-digit number has a digit sum of one. How many zero digits does the number have?

Exercise 15

Dina has written a 3 digit number on a piece of paper. Lila must guess which number it is. Dina gives Lila a hint: "One of the digits is 5. The number does not change if you move the last digit in front of the first." Which number is it?

Exercise 16

Compute:

(a) $10 - 9 + 8 - 7 + 6 - 5 + 4 - 3 + 2 - 1 + 0 =$

(b) $100 + 99 - 99 + 98 - 98 + 97 - 97 + 96 - 96 + 95 - 95 + 94 - 94 + 93 - 93 + 92 - 92 + 91 - 91 + 1 =$

(c) $9 - 2 + 3 - 4 + 5 =$

Exercise 17

Lila has 5 more toys than Amira does. If Lila gives Amira 7 toys, how many more toys than Lila will Amira have?

Exercise 18

Find the positive integers that are hidden by symbols. Different symbols hide different integers. How many different solutions can be found?

$$\Diamond + \heartsuit + \heartsuit = 7$$

Exercise 19

In the following sequence of consecutive odd numbers, how many numbers have been replaced by dots?

$$15, \ 17, \ \cdots , \ 25$$

Exercise 20

How many numbers between 200 and 240 can be written using only the digits 2 and 3?

Exercise 21

A machine crunches numbers and outputs the result. The figure shows a set of numbers entering the machine and the results of the crunches coming out of the machine. If the machine is given the number 207 to crunch, what will it output?

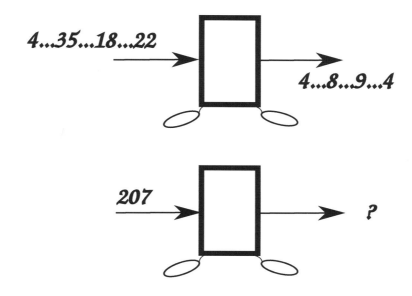

Exercise 22

Another machine mashes numbers and outputs the result. The figure shows a set of numbers entering the machine and the results of the mashes coming out of the machine. If the machine is given the number 207 to mash, what will it output?

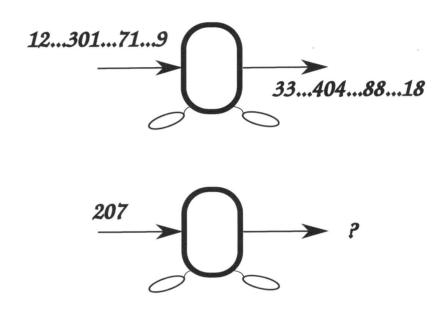

SOLUTIONS TO PRACTICE ONE

Solution 1

$$16 + 0 = 16$$

Exercise 2

True or false?

$$102 - 0 = 102 + 50 - 50$$

Solution 2

Since $50 - 50$ equals zero, the left hand side is equal to the right hand side. True.

Exercise 3

True or false?

$$102 + 0 = 102 - 50 + 50$$

Solution 3

This exercise is similar to the one above, except we now subtract 50 *before* adding it back in. Subtracting and then adding the same quantity from the original number does not change it, so the result is still 102. True.

Exercise 4

True or false?

$$5 + 2 - 2 = 2 - 2 + 5$$

Solution 4

True. Adding and then subtracting 2 from a number leaves the number unchanged.

The right hand side models the following situation: "Dina places two one dollar bills on the table. She then removes them and places a five dollar bill on the table. How many dollars are there on the table now?"

The left hand side models the following situation: "Dina places a five dollar bill on the table. She then places two one dollar bills on the table. She then removes the two one dollar bills. How many dollars are there on the table now?"

Exercise 5

Dina computed the following:

$$100 + 11 - 11 + 12 - 12 + 13 - 13 =$$

$$100 - 11 + 11 - 12 + 12 - 13 + 13 =$$

Solution 5

Dina noticed that she gets the same result:

$$100 + 11 - 11 + 12 - 12 + 13 - 13 = 100$$

$$100 - 11 + 11 - 12 + 12 - 13 + 13 = 100$$

In both cases, we add and then subtract an 11. The same is true for 12 and 13 and, therefore, the initial number (100) remains unchanged.

Dina said: "If you add and subtract the same number from another, the initial number does not change. It does not make a difference if you add first and then subtract or the other way around."

Lila said: "Wow! Then we don't actually need to do any computations. We just have to notice how many add-subtract pairs of the same number there are and cross them out!"

Exercise 6

$$100 + 99 + 98 + 97 + 96 + 95 + 94 - 95 - 96 - 97 - 98 - 99 - 100 =$$

Solution 6

Lila noticed that there were a lot of add-subtract pairs:

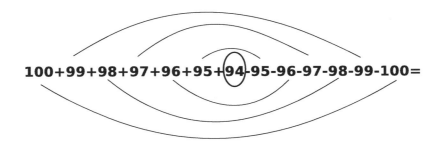

She crossed out the instances where the same number was both added and subtracted and was left with the number 94.

Solution 7

43

Solution 8

9

Exercise 9

Dina performed these additions and noticed something interesting. What did Dina notice?

$$123 + 321 = 444$$
$$342 + 243 = 545$$
$$241 + 142 = 383$$
$$721 + 127 = 848$$

Solution 9

Dina noticed that each of the results is a palindrome (a number that remains unchanged when the order of its digits is reversed).

Exercise 10

Place the numbers 15, 12, and 11 in the squares to make the following equality true:

$$\square \ + \ \square \ - \ \square \ = \ \mathbf{16}$$

Solution 10

A possible solution is:

$$\boxed{12} \ + \ \boxed{15} \ - \ \boxed{11} \ = \ 16$$

Another solution is possible if we use the commutative property of addition.

Exercise 11

If Dina wants to add 5 and 4 and then subtract 2 from the sum, what is she going to have to enter?

(A) $+$ 4 $5-2$

(B) $+$ $-$ 5 4 2

(C) $-$ $+$ 4 5 2

Solution 11

If Dina enters choice (A), the calculator will complain like this: "SYNTAX ERROR." This means that it does not understand the input. While $+$ 4   5 produces the answer 9, the calculator does not know what to do with the $9 - 2$ input since it expects the operator to be *in front of* the numbers it applies to.

If Dina enters choice (B), the calculator will find it knows how to compute $- 5$ 4. It will use the result (1) in the next operation. After this, the calculator will see the input $+ 1$ 5. It can understand this, since the operator is in front of the two numbers. It will produce the answer 6.

If Dina enters choice (C), the calculator will find it knows how to compute $+ 4$ 5 and it will get 9. After this is done, it will see the new input, $- 9$ 2, which it knows how to compute. It will get the answer 7.

The correct answer is (C).

Solution 12

The calculator reads the input from left to right and immediately performs the operations it understands. It replaces the operation with its result and reads the input again from left to right, like in this diagram:

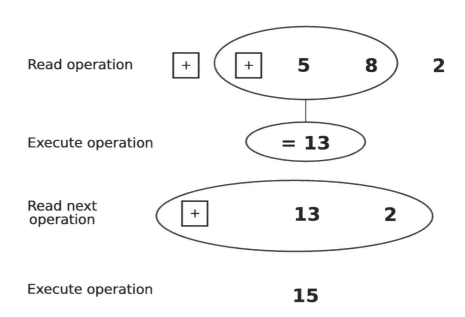

(a) $+$ 7 8 $=$ 15

(b) $+$ $+$ 7 8 2 $=$ $+$ 15 2 $=$ 17

(c) $-$ $+$ 7 8 2 $=$ $-$ 15 2 $=$ 13

(d) $+$ 5 3 $=$ 8

(e) $-$ 5 3 $=$ 2

(f) $\times$ 5 2 $=$ 10

(g) $\div$ $\times$ 5 2 2 $=$ $\div$ 10 2 $=$ 5

(h) $\div$ $+$ 5 5 2 $=$ $\div$ 10 2 $=$ 5

(i) $\times$ $-$ 5 5 4 $=$ $\times$ 0 4 $=$ 0

(j) $+$ $+$ $+$ 3 3 3 3 $=$ $+$ $+$ 6 3 3 $=$ $+$ 9 3 $=$ 12

(k) $-$ $-$ $-$ 4 1 1 1 $=$ $-$ $-$ 3 1 1 $=$ $-$ 2 1 $=$ 1

(l) $-$ $+$ 4 3 2 = $-$ 7 2 = 5

(m) $-$ $+$ 1 9 1 = $-$ 10 1 = 9

Exercise 13

Fill in the missing values.

Solution 13

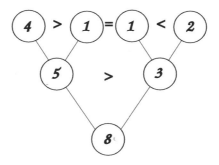

Exercise 14

Dina and Lila have entered the following operations in their calculators. Each calculator performs one operation at a time. Fill in the blanks with the numbers each calculator used for each step.

Solution 14

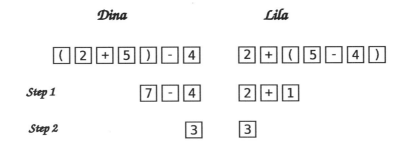

Exercise 15

Find a pattern and fill in the missing values.

Solution 15

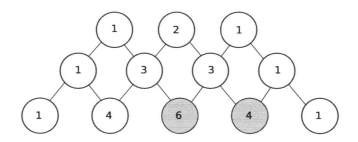

Exercise 16

Find the number in the grey circle.

Solution 16

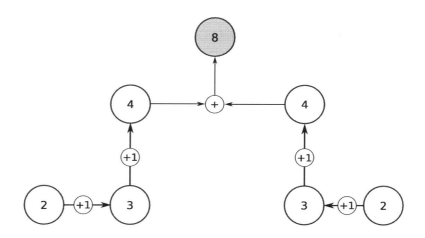

Exercise 17

Place the results of the operations in the boxes such that identical results go in the same box. How many boxes have remained empty?

Solution 17

The results are:

$$
\begin{aligned}
95 + 10 &= 105 \\
15 - 0 &= 15 \\
100 - 85 &= 15 \\
220 - 105 &= 115 \\
77 + 38 &= 115 \\
8 + 7 &= 15 \\
45 + 60 &= 105 \\
30 - 15 &= 15 \\
121 - 91 &= 30 \\
67 - 37 &= 30
\end{aligned}
$$

There are only 4 distinct results. Therefore, one box remains empty.

SOLUTIONS TO PRACTICE TWO

Exercise 1

True or false?

$$1 - 2 + 1 = 1 + 1 - 2$$

Solution 1

On both sides, we add 1 twice and subtract 2. The result is zero on both sides.
True.

Exercise 2

True or false?

$$3 - 4 = 4 - 3$$

Solution 2

On the left hand side, we add 3 and subtract 4: the result is -1.
On the right hand side, we add 4 and subtract 3: the result is 1.
False.

Exercise 3

True or false?

$$3 - 4 = -4 + 3$$

Solution 3

On the left hand side, we add 3 and subtract 4: the result is -1.
On the right hand side, we add 3 and subtract 4: the result is -1.
True.

Exercise 4

Dina and Lila play a game on the number line. If Dina says a positive number, Lila will move that number of steps to the right. If Dina says a negative number, Lila will move that number of steps to the left. When Dina does not say anything, Lila stays put. Dina and Lila are facing each other. Dina says:

$$-2, \ 3, \ 1, \ -4, \ 2, \ -1, \ 3, \ -2, \ 1, \ 1, \ -4$$

Is Lila now to Dina's left or to Dina's right? How many steps away from Dina is Lila now?

Solution 4

Make a model of a number line and move the tip of the pencil left and right from the starting position.

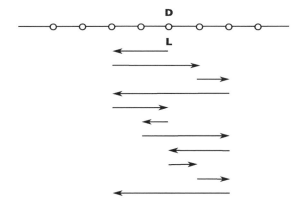

Lila ends up two steps away from Dina's right.
This result can also be obtained by performing the operations:

$$-2 + 3 + 1 - 4 + 2 - 1 + 3 - 2 + 1 + 1 - 4 = -2$$

Exercise 5

Lila must place the operators $+$ and $-$ in the empty squares so that the following equality is true. Can you help her?

Solution 5

Exercise 6

Compute efficiently:

Solution 6

(a) $22 - 5 + 1 - 8 = 10$

Add the positive numbers together and the negative numbers together: $22 + 1 = 23$ and $5 + 8 = 13$. Now, the operation is:

$$23 - 13 = 10$$

(b) $101 + 59 - 100 = 60$

Subtract 100 from 101 and get 1. Add 1 to 59 to get 60.

(c) $119 + 1 - 49 - 1 = 70$

Cross out the $+1$ -1 pair. You are left with:

$$119 - 49 = 70$$

(d) $77 - 2 - 3 - 4 - 5 - 6 = 57$

Add together all the numbers that have to be subtracted from 77. Add 4 and 6 to get 10. Add 2, 3, and 5 to get 10. All the numbers to be subtracted add up to 20. Subtract 20 from 77 to get 57.

(e) $299 + 1 - 299 = 1$

Notice that you are adding and then subtracting 299. Cross out both terms. You are left with 1.

Exercise 7

True or false?

Solution 7

(a) $11 - 4 - 5 - 6 - 7 + 99 = 110 - (4 + 5 + 6 + 7)$

Add together the positive values on both sides of the equation, then add together the negative numbers. On both sides, 11 and 99 are added and 4, 5, 6, and 7 are subtracted. The equality is true.

(b) $20 - 30 + 10 = 20 + 10 - 30$

True.

(c) $2 - 3 + 4 - 5 + 6 = 6 - 5 + 4 - 3 + 2$

On both sides, 2, 4, and 6 are added and 3 and 5 are subtracted. Both sides produce the same result. True.

(d) $(13 + 12) - (11 + 14) = 12 - 11 + 13 - 14$

Notice how, on the right hand side, 12 and 13 are added and 11 and 14 are subtracted. This is the same as subtracting the sum of 11 and 14 from the sum of 13 and 12. The equality is true.

Exercise 8

True or false?

Solution 8

(a) $23 + 25 + 27 = 28 - 1 + 26 - 1 + 24 - 1$

Notice that $28 - 1 = 27$, $26 - 1 = 25$, and $24 - 1 = 23$.

Both sides are the same. True.

(b) $16 + (1 - 16) = (16 - 16) + 1$

On both sides, we add a 16 and then subtract a 16 (cross out both). Both sides are equal to 1. True.

(c) $11 - (5 + 6) = 11 - 5 + 6$

On the left hand side, we subtract 5 and 6 from 11. The left hand side is equal to 0. On the right hand side we subtract 5 from 11 and then add 6. The right hand side is equal to 12. False.

(d) $5 + 15 + 25 - 20 - 10 = 5 + (15 - 10) + (25 - 20)$

On both sides, we add 5, 15, and 25 and then subtract 10 and 20. True.

Exercise 9

Lila has to place parentheses to obtain the smallest possible result.

$$43 \quad - \quad 3 \quad + \quad 15 \quad - \quad 7$$

Solution 9

We want to subtract the largest number possible:

$$43 - (3 + 15) - 7 = 43 - 18 - 7 = 18$$

Exercise 10

Dina has to place parentheses to obtain the result shown:

$$49 \quad - \quad 8 \quad - \quad 41 \quad + \quad 37 \quad = \quad 45$$

Solution 10

Dina notices that $37 + 8 - 41 = 4$ and that by subtracting 4 from 49 she obtains the required result. She places parentheses so that the numbers that add up to 4 are grouped together.

$$49 - (8 - 41 + 37) = 45$$

Exercise 11

Compute the result:

$$2 + 4 + 6 + 8 + 10 - 1 - 3 - 5 - 7 - 9 =$$

Solution 11

$$2 + 4 + 6 + 8 + 10 - 1 - 3 - 5 - 7 - 9 = 5$$

Notice that:

$$10 - 9 = 1$$
$$8 - 7 = 1$$
$$6 - 5 = 1$$
$$4 - 3 = 1$$
$$2 - 1 = 1$$

Exercise 12

Dina has to place parentheses in the following expression so that the computation is correct:

Solution 12

$$(40 + 8) \div 8 = 6$$

Exercise 13

Lila has to place parentheses in the following expression so that the computation is correct:

Solution 13

$$(9 + 8) \times 3 = 51$$

Exercise 14

Place parentheses in the following expression so that the computation is correct:

Solution 14

$$(8 + 10) \times 2 = 36$$

Exercise 15

Place parentheses in the following expression so that the computation is correct:

Solution 15

$$(4+5) \times (9-3) = 9 \times 6 = 54$$

Exercise 16

Without calculating, can you tell whether each line is true or false?

Solution 16

$10 - 1 + 11 - 1 = 10 + 11 - 2$	True
$8 - 3 = -3 + 8$	True
$8 - 3 = 3 - 8$	False
$5 + 3 - 3 = 8 + 5 - 8$	True
$11 - 12 + 12 - 1 = 11 - 1$	True

Exercise 17

True or false?

Solution 17

Pair the identical numbers that are both added and subtracted. Both sides are equal to zero.

$$2 - 2 + 3 - 3 + 4 - 4 + 5 - 5 = -2 + 2 - 3 + 3 - 4 + 4 - 4 + 5$$

True.

Exercise 18

True or false?

Solution 18

On both sides, we see pairs of numbers that are add-subtract pairs. Both sides are equal to zero.

$$5+6+7+8+9-5-6-7-8-9 = 5+6+7+8+9-(5+6+7+8+9)$$

Solutions to Practice Three

Exercise 1

$$\{2,\ 4,\ 6,\ 8,\ 10,\ \cdots, 20\}$$

(a) How many numbers are written out?

(b) How many numbers have been replaced by dots?

(c) How many numbers are there in total (written and unwritten)?

Solution 1

The complete list is $\{2,\ 4,\ 6,\ 8,\ 10,\ 12,\ 14,\ 16,\ 18,\ 20\}$.

(a) 6 numbers are written out in the shortened list.

(b) 4 numbers have been replaced by dots.

(c) The list contains 10 numbers in total.

Exercise 2

$$\{2,\ 4,\ 6,\ 8,\ 10,\ \cdots, 40\}$$

(a) How many numbers are written out?

(b) How many numbers have been replaced by dots?

(c) How many numbers are there in total (written and unwritten)?

Solution 2

This problem has more numbers than the previous problem to show students how important it is to make the transition from brute force to strategic reasoning.

Brute force solution:

The complete list is:

$$\{2, 4, 6, 8, 10, 12, 14, 16, 18, 20, 22, 24, 26, 28, 30, 32, 34, 36, 38, 40\}$$

(a) 6 numbers are written out.

(b) 14 numbers have been replaced by dots.

(c) The list contains 20 numbers in total.

Strategic solution:

Notice that the list is made up of consecutive even numbers. Figure out how many even numbers there are from 1 to 40. Half of them (20) are even. This is how many numbers there are in the list that is written out completely. In the abbreviated list, only 6 numbers are written out. $20 - 6 = 14$ numbers have been replaced by dots.

Exercise 3

Can you find out how many numbers have not been written out?

$$\{2,\ 4,\ 6,\ 8,\ 10,\ \ldots, 400\}$$

Solution 3

There must be 200 numbers in total. 6 numbers have been written out. The remaining $200 - 6 = 194$ numbers have been replaced by dots.

Exercise 4

Dina has to answer some questions about the following expression:

$$1 + 2 + 3 + \cdots + 40$$

1. How many numbers are added together?
2. How many of these numbers are odd?
3. How many + operators are there in total?
4. Will the result be even or odd? (Answer without calculating the result.)

Solution 4

1. 40 numbers have been added together.

2. Half of them (20 numbers) are odd.

3. Operators only occur between numbers (remember what you learned in "Practice Counting"). There are 39 plus signs.

4. Since there are 20 odd terms, they can be paired to form even sums. All the even numbers will add up to an even number. The final result will be even.

Solution 5

3 squares are the same as (equivalent to) one triangle.

$$\square \ + \ \square \ + \ \square \ + \ \square \ = \ \triangle \ + \ \square$$

Solution 6

100 squares are the same as (equivalent to) one star.

$$\underbrace{\square \ + \ \square \ + \ \cdots \ + \ \square}_{101 \text{ squares}} \ = \ \bigstar \ + \ \square$$

Solution 7

22 squares are the same as (equivalent to) one circle.

$$\underbrace{\square \ + \ \square \ + \ \cdots \ + \ \square}_{66 \text{ squares}} \ = \ \bigcirc \ + \ \bigcirc \ + \ \bigcirc$$

Exercise 8

Dina has to count the number of operations in the following sum:

$$1 + 2 + 3 + 4 + 5 + 6 + 7 + 8 + 9 + 10$$

Solution 8

There are 10 numbers to add. Therefore, 9 additions (operations) are necessary.

Exercise 9

Lila has to count the number of operations in the following sum:

$$1 + 2 + 3 + 4 + 5 + \cdots + 100$$

If Lila knows what Dina found in the previous problem, can she find an answer without writing out all the numbers?

Solution 9

There are 100 numbers to add. Therefore, 99 additions (operations) are necessary.

$$1 + 2 + 3 + 4 + 5 + \cdots + 100$$

Exercise 10

Compute efficiently:

$$2 + 4 + 6 + 8 + 10 + \cdots + 50 - 1 - 3 - 5 - 7 - 9 - \cdots - 49 =$$

Solution 10

Pair the terms in a different way:

$$2 - 1 + 4 - 3 + 6 - 5 + 8 - 7 + 10 - 9 + \cdots + 50 - 49 =$$

and notice how each pair is equal to 1.

Since there are 50 numbers in total, there must be 25 pairs.

The sum of the numbers in each pair is equal to 1, so the total is 25.

Exercise 11

How many numbers are there in the list?

$$\{11,\ 22,\ \cdots,99,\ 111,\ 222,\ \cdots,999\}$$

Solution 11

There are nine 2-digit numbers and nine 3-digit numbers.
There are 18 numbers in total.

Exercise 12

How many numbers are there in the list?

$$\{1,\ 11,\ 111,\ 1111,\ \cdots,111111111111111\}$$

Solution 12

Each number has one more digit than the preceding one. Since the last number has 15 digits, there are 15 numbers in the list.

Exercise 13

How many numbers in the list are even?

$$\{0,\ 1,\ 2,\ 3,\ \cdots,100\}$$

Solution 13

From 1 to 100 there are 50 even numbers. Since zero is also even, the total number of even numbers in the list is 51.

Exercise 14

How many towers are there?

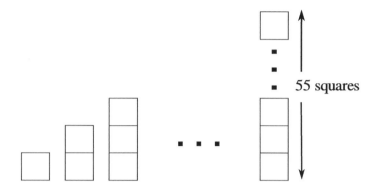

Solution 14

Each tower is one block higher than the previous tower. There are 55 towers in total.

Exercise 15

How many circles are filled?

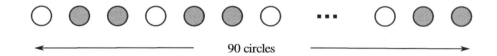

Solution 15

The circles can be placed in groups of three, with one white and two shaded circles in each group.

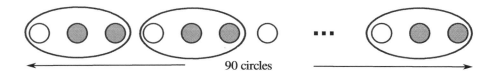

There are 30 groups of three in 90. In each group of three, two circles are shaded. The total number of shaded circles is 60.

Exercise 16

How many digits have been used to write the list:

$$\{1, \ 22, \ 333, \ \cdots, 7777777, \ \cdots, 333, \ 22, \ 1\}$$

Solution 16

The number of digits used is equal to the sum:

$$1 + 2 + 3 + 4 + 5 + 6 + 7 + 6 + 5 + 4 + 3 + 2 + 1 =$$

Since:

$$1 + 2 + 3 + 4 + 5 + 6 = 21$$

We have this sum twice plus a 7. The total is:

$$21 + 21 + 7 = 49$$

49 digits have been used.

Note: The problem did not ask for the number of *different (distinct)* digits. If that had been the case, the answer would have been 7.

Exercise 17

Lila and Dina are playing a game of dots. They use a ruled sheet of paper. Dina starts by drawing a dot on the first line. Lila draws two dots on the second line. Dina draws three dots on the third line. They continue on until one of them draws 47 dots on a line. Which line is it? Who draws the 47 dots, Lila or Dina?

Solution 17

The number of dots on each line is the same as the line number. Since Lila and Dina take turns drawing the dots, Lila will draw the dots on all the even numbered lines and Dina will draw the dots on all the odd numbered lines. Therefore, Dina will draw 47 dots on the 47th line.

SOLUTIONS TO PRACTICE FOUR

Exercise 1

Which number is hiding behind the square?

Solution 1

20

Exercise 2

Which number is hiding behind the circle?

Solution 2

11

Exercise 3

Fill in the circles with appropriate integer numbers:

Solution 3

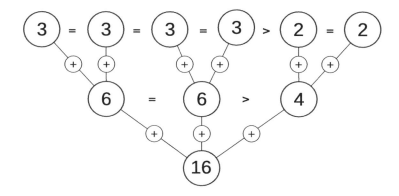

Exercise 4

Find K:

$$61 + 59 = K + K + K$$

Solution 4

$$
\begin{aligned}
120 &= K + K + K \\
120 &= 30 + 30 + 30 \\
K &= 30
\end{aligned}
$$

Exercise 5

Place the same number in each empty circle:

$$\bigcirc \times \bigcirc - \bigcirc / \bigcirc = 3$$

Solution 5

$$2 \times 2 - 2 / 2 = 3$$

Exercise 6

Place in the circles numbers formed using the same digit:

$$\bigcirc \times \bigcirc - \bigcirc / \bigcirc = 88$$

Solution 6

$$33 \times 3 - 33 / 3 = 88$$

Exercise 7

Place operators $(+, -, \times, \div)$ within the boxes. No parentheses are needed.

Solution 7

Exercise 8

In each of the following, which number does the letter represent?

$A + A + A = 15$

$B + B + B + B - B = 15$

$C + C + C + C + C = 15$

$C + C + C + C + C + C - C = 15$

$D + D - D + D - D + D - D + D - D + D - D = 15$

Solution 8

Use the fact that repeated addition is multiplication. Also, adding and subtracting the same number from a number leaves it unchanged. Cross-out add/subtract pairs of identical numbers.

$$A = 5$$
$$B = 5$$
$$C = 3$$
$$C = 5$$
$$D = 15$$

Exercise 9

The triangle, the circle, and the square have different weights. Which of the following would balance the scale on the right?

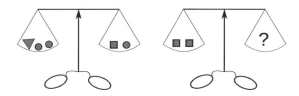

Solution 9

The balance on the left contains one triangle and two circles in one pan, and one square and one circle in the other pan. Since the two pans are perfectly balanced, one square must weigh as much as one triangle and one circle. Therefore, two squares will have the same weight as two triangles and two circles. The correct answer is (D).

Exercise 10

Amira is out shopping for Hallowe'en treats. She wants to buy some Real Fruit Eyeballs and some Fruit Jelly Smellies. Three packs of Eyeballs and two packs of Smellies cost 9 dollars. Two packs of Eyeballs and three packs of Smellies cost 11 dollars. Amira has 12 dollars. Does she have enough money to buy three packs of Eyeballs and three packs of Smellies?

Solution 10

Five packs of Eyeballs and five packs of Smellies cost $9 + 11 = 20$ dollars. Therefore, one pack of Eyeballs and one pack of Smellies cost 4 dollars and two packs of Eyeballs and two packs of Smellies cost 8 dollars. Since we know that two packs of Eyeballs and three packs of Smellies cost 11 dollars, one pack of Smellies must cost 3 dollars. This means that a pack of Eyeballs costs one dollar. Three packs of Eyeballs and three packs of Smellies cost exactly 12 dollars. Amira can make her purchase!

SOLUTIONS TO MISCELLANEOUS PRACTICE

Exercise 1

Which number does the circle represent?

$$\bigcirc + \bigcirc = \star$$

$$\star + \star = 136$$

Solution 1

Each circle is the same as half a star. Each star represents half of 136. Therefore, the star represents 68 and the circle represents 34.

Exercise 2

Replace the question mark with a number so that the operations are correct:

$$\bigcirc + \star = 16$$

$$\bigcirc - \star = 4$$

$$\bigcirc + \bigcirc = \ ?$$

Solution 2

In the first operation, the star is added to the circle and in the second one, the star is subtracted from the circle. Therefore, $16 + 4$ equals two circles, plus a star, minus a star. Adding and subtracting the same amount does not change anything. Cross-out the add-subtract pair of the same symbol. Two circles equal 20.

Exercise 3

If two positive integers have an odd difference, is their sum:

(**A**) always even?

(**B**) always odd?

(**C**) sometimes even and sometimes odd?

Solution 3

If their difference is odd, then the numbers must have different parity. One must be even, while the other must be odd. Therefore, their sum must also be odd. The answer is (B).

Exercise 4

There are 100 plus signs in the following operation. What is the result of the additions?

$$1+1+1+\cdots+1= \text{?}$$

Solution 4

If there are 100 operators, then there must be 101 terms, each equal to 1. The sum is equal to 101.

Solution 5

Roman numerals time! Perform the following operations and write the answer in both Roman and Arabic numerals:

(a) $I + I = II$ $(1 + 1 = 2)$

(b) $II + II = IV$ $(2 + 2 = 4)$

(c) $III + III = VI$ $(3 + 3 = 6)$

(d) $IV + IV = VIII$ $(4 + 4 = 8)$

(e) $V + V = X$ $(5 + 5 = 10)$

(f) $XX + XX = XL$ $(20 + 20 = 40)$

(g) $XXX + XXX = LX$ $(30 + 30 = 60)$

(h) $LX + LX = CXX$ $(60 + 60 = 120)$

(i) $L + L = C$ $(50 + 50 = 100)$

(j) $C + C = CC$ $(100 + 100 = 200)$

(k) $CC + CC = CD$ $(200 + 200 = 400)$

(l) $CCC + CCC = DC$ $(300 + 300 = 600)$

(m) $CD + CD = DCCC$ $(400 + 400 = 800)$

(n) $CD + D = CM$ $(400 + 500 = 900)$

Exercise 6

Each group of four numbers in the sequence follow the same pattern. Which numbers belong in the empty circles?

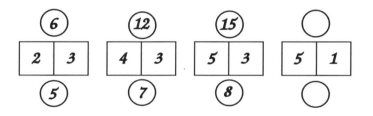

Solution 6

The circle on top contains the product of the numbers in the squares. The circle on the bottom contains the sum of the numbers in the squares:

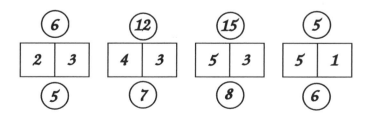

74

Exercise 7

In the figure, shaded circles of the same color hide the same number. Which number does the white circle hide?

Solution 7

The sum of the six white circles is 12. Therefore, each white circle must hide a 2. The intermediate values can be determined but are not necessary in deriving the answer.

Exercise 8

Compute:

$$4 + 3 - 4 + 5 - 6 + 6 - 7 + 8 - 9 =$$

Solution 8

Notice how the terms can be grouped in pairs that are equivalent to subtracting 1. Each of the operations $3 - 4$, $5 - 6$, $6 - 7$, and $8 - 9$ subtracts a 1. We have to subtract 1 four times from the first term. The result is zero.

Exercise 9

Dina had 30 party balloons. During the party, 11 balloons popped. Afterwards, Lila gave Dina as many balloons as had popped. How many party balloons did Dina have then?

Solution 9

In this problem, we subtract a number and then add it back. These operations do not change the initial number. Dina had 30 balloons after Lila gave her as many balloons as had popped.

Exercise 10

The table in the figure has 10 rows and 11 columns. How many grey cells are there in the table? How many grey cells should we color white in order to have an equal number of white and grey cells?

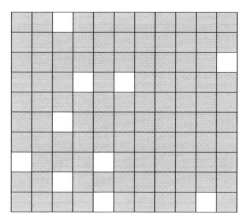

Solution 10

Notice that there are 10 white squares. The table has 10 rows and 11 columns, therefore it must have 110 cells. Subtract $110 - 10 = 100$ to obtain the number of grey cells.

There are 90 more grey cells than white cells. In order to have an equal number of white and grey cells, we need to paint 45 cells white.

Exercise 11

In the lists below, some digits, not necessarily identical, have been hidden by a ♠. In each list, there are two different 3-digit numbers. Find the lists in which it is possible to tell which number is smaller.

(A) 9♠7, ♠87

(B) 33♠, 31♠

(C) 4♠3, 5♠♠

(D) 9♠♠, 9♠0

Solution 11

Only in lists (B) and (C) is it possible to tell which number is smaller without knowing which digits are hidden by the ♠s.

Exercise 12

Dina has 20 magnets and Lila has 11 magnets. How many magnets could Amira have if she has fewer than Dina and more than Lila? Check all that apply.

(A) 9

(B) 10

(C) 13

(D) 19

Solution 12

The number of magnets Lila has must be larger than 11 and smaller than 20. Possible numbers are: 12, 13, 14, 15, 16, 17, 18, and 19. Answer choices (C) and (D) are in this list.

Exercise 13

How many 3-digit numbers have a digit sum of 2?

Solution 13

Three numbers: 110, 101, and 200.

Exercise 14

A 100-digit number has a digit sum of one. How many digits 0 does the number have?

Solution 14

Since zero cannot be the first digit of a number, the number must start with a digit of 1 or greater. Because the digit sum must be 1, then the number must start with a 1 and all the remaining digits must be zero. The 1 is followed by 99 zeros.

Exercise 15

Dina has written a 3 digit number on a piece of paper. Lila must guess which number it is. Dina gives Lila a hint: "One of the digits is 5. The number does not change if you move the last digit in front of the first." Which number is it?

Solution 15

By moving the last digit to the leftmost position, the middle digit becomes the last. For the number to remain unchanged, therefore, the middle digit must equal the last digit. Also, the leftmost digit becomes the middle digit and, therefore, the leftmost digit must equal the middle digit. As a result, all the digits must be identical. The number must be 555.

Solution 16

Compute:

(a) $10 - 9 + 8 - 7 + 6 - 5 + 4 - 3 + 2 - 1 + 0 = 1 + 1 + 1 + 1 + 1 = 5$

(b) $100 + 99 - 99 + 98 - 98 + 97 - 97 + 96 - 96 + 95 - 95 + 94 - 94 + 93 - 93 + 92 - 92 + 91 - 91 + 1 = 101$

(c) $9 - 2 + 3 - 4 + 5 = 9 + 1 + 1 = 11$

Exercise 17

Lila has 5 more toys than Amira. If Lila gives Amira 7 toys, how many more toys than Lila will Amira have?

Solution 17

Amira would have 9 more toys than Lila:

Before:

Lila's toys [| 5]
Amira's toys [   ]

Lila gives 7 toys to Amira:

Lila's toys [|2| 5]
Amira's toys [|2| 5]

After:

Lila's toys [   ]
Amira's toys [|2|2| 5]

Exercise 18

Find the positive integers that are hidden by symbols. Different symbols correspond to different integers. How many different solutions can be found?

$$\diamondsuit + \heartsuit + \heartsuit = 7$$

Solution 18

3 solutions can be found:

$$1 + 3 + 3 = 7 \qquad 3 + 2 + 2 = 7 \qquad 5 + 1 + 1 = 7$$

Exercise 19

In the following sequence of consecutive odd numbers, how many numbers have been replaced by dots?

$$15, \ 17, \ \cdots, 25$$

Solution 19

There are 3 numbers missing: 19, 21, and 23.

Exercise 20

How many numbers between 200 and 240 can be written using only the digits 2 and 3?

Solution 20

Four numbers: 222, 223, 232, and 233.

Exercise 21

A machine crunches numbers and outputs the result. The figure shows a set of numbers entering the machine and the results of the crunches coming out of the machine. If the machine is given the number 207 to crunch, what will it output?

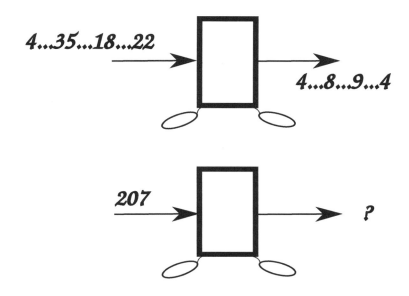

Solution 21

Notice that the machine outputs the digit sum of the numbers that it crunches. Therefore, if the input is 207, the machine will output 9.

Exercise 22

Another machine mashes numbers and outputs the result. The figure shows a set of numbers entering the machine and the results of the mashes coming out of the machine. If the machine is given the number 207 to mash, what will it output?

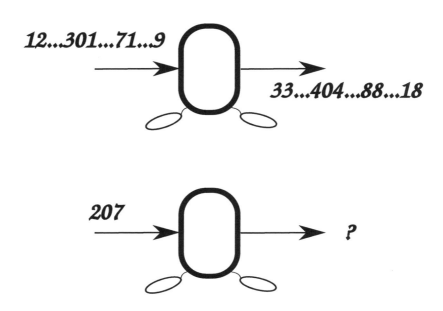

Solution 22

Notice that the machine reverses the digits of the number and then adds the new number to the initial one:

$$
\begin{aligned}
12 + 21 &= 33 \\
301 + 103 &= 404 \\
71 + 17 &= 88 \\
9 + 9 &= 18
\end{aligned}
$$

Therefore, if the input is 207, the machine will output $207 + 702 = 909$.

Competitive Mathematics Series for Gifted Students

Practice Counting (ages 7 to 9)
Practice Logic and Observation (ages 7 to 9)
Practice Arithmetic (ages 7 to 9)
Practice Operations (ages 7 to 9)

Practice Word Problems (ages 9 to 11)
Practice Counting (ages 9 to 11)
Practice Arithmetic (ages 9 to 11)
Practice Operations (ages 9 to 11)

Practice Word Problems (ages 11 to 13)
Practice Arithmetic (ages 11 to 13)
Practice Operations and Algebra (ages 11 to 13)
Practice Geometry (ages 11 to 13)

Practice Word Problems (ages 12 to 15)
Practice Operations (ages 12 to 15)
Practice Geometry (ages 12 to 15)
Practice Algebra (ages 12 to 15)

This is a series of practice books. With the exception of a few reminders, there are no theoretical explanations. For lessons, please see the resources indicated below:

Find a set of free lessons in competitive mathematics at www.mathinee.com. Addressing grades 5 through 11, the *Math Essentials* on www.mathinee.com present important concepts in a clear and concise manner and provide tips on their application. The site also hosts over 400 original problems with full solutions for various levels. Selectors enable the user to sort essentials and problems by test or contest targeted as well as by topic and by the earliest grade level they can be used for.

Online problem solving seminars are available at www.goodsofthemind.com. If you found this booklet useful, you will love the live problem solving seminars.